I0463477

Reclaiming the Universe

By Zak Ettamymy

Table of Contents

Zak Ettamymy

Preface

The universe is full of questions, contradictions, peculiarities, heresies and enigmas, many great minds tried to tackle them but failed. It's understandably difficult, after all if you are inside the experiment it's hard to observe, analyze and judge the same experiment.

One thing is sure, mankind made many strides in proving to itself (since humans are the only known conscious entity in the universe) that it can contemplate the big question, the question that only a thinking species can dare to ask: What is the origin of all of this? I doubt that animals spend their lives pondering the meaning of this universe; I doubt that dinosaurs and ancient plants had the urge to know where they came from millions of years ago. But we do and we did since we started to think about 30 thousand years ago. But what does this all mean? Doesn't the universe mean anything to these poor unthinking species? Do they know that they're alive

Zak Ettamymy

and that they follow certain laws of the universe? Do they feel that they have a special time and place in it? Hard to know but one thing is sure we are aware of our place and time in the universe and that is what makes this universe ours, because we thought about it, mapped it, gave it a name and age, and we even messed around with its laws, we made up reasons and scenarios for its birth and we produced elaborate schemes to convince each other of the results.

What can we do with the information we collected so far? What are the implications this information has on our quest to spirituality and the relationship between matter and soul?

A Spiral Galaxy in a multi galactic background

History of Humans and their Universe

Humans were always fascinated by the vast "world" surrounding them. They knew something was special about the sky above for centuries, although most records about Man's contemplation of the heavens were made during the antiquity, humans started long before that with simple observations and some record keeping of the movements of the stars and the planets, which they differentiated due to their movement of lack of it. The notion of gods and spirits up in the sky watching down these primitive people who roamed the planet was the widely accepted notion regarding the celestial relationship with them, They worshiped the Moon, the Sun and the Stars, later they started to see them as floating objects up in the sky that had no significant power over them but the curiosity kept Man looking up, trying to find a place for the heavens in his primitive life.

The Neanderthals and other primitive humans, although lived almost at the same time as the early homo sapiens, never had a religion, never had deep thoughts about the meaning of life, although they fought to preserve it yet failed (they were extinct eventually) to contemplate its purpose, they never looked up to the sky with an investigative eye, they never thought of the universe as theirs, Homo sapiens however saw themselves as part of it, they worshiped it and interacted with it. This is where conscientiousness started to play a role in the universe.

The discussion about the meaning of existence as far as awareness is concerned could lead us to philosophical realms, but in the early years of modern humans, they made the distinction between survival existence that the Neanderthals had and the deeper meaning of existence that we modern humans express until now.

Zak Ettamymy

An ancient depiction of the skies

The bronze age came with promising discoveries by introducing the first calendar about ten thousand years ago in ancient Europe, as well as the kingdoms of ancient Iraq (Samaria, Assyria and Babylonia) who were the first to use mathematics to chart and define astronomical laws and patterns, beautifully written in a collection of tablets called Enuma Anu Enlil, they used sand and water clocks to set the recurrence of the sun set and sun rise and the passing of the seasons and years.

The Babylonia astronomy was the basis of all future astronomies, such as the Greek, the Indian, the Islamic and Western Europe. All used the first interpretation of the Babylonian astronomers and priests to enhance and expand astronomy in their prospective culture, except ancient Egyptians, who it is believed to have had a slightly different view on astronomy. For the first time, in humans' history, the Egyptians started using tools and handmade measuring devices to explain and chart astronomical phenomenon and sometimes to answer questions that took thousands of years to be proven true, the jewel of which is the fact that earth is not flat.

Babylonian tablet recording Halley's Comet in 164 BC

In the Library of Alexandria Worked Eratosthenes of Cyrene, a man with many talents and one of them was the sense of logic or analytic thinking rather than the widely used dogmatic interpretations of the time. One day Eratosthenes received some amazing correspondence from the city of Syene in southern Egypt. In particular, it said that, on the Summer Solstice, *the shadow of someone looking down a deep well would block the reflection of the Sun at noon.* In other words, the Sun would be directly overhead at this time, not a single degree to the South, North, East or West. And if you had a completely vertical object, it would cast *absolutely no shadow*. But Eratosthenes knew that this *wasn't* the case where he was, in Alexandria. Sure, the Sun came **closer** to being directly overhead at Noon on the Summer Solstice in Alexandria than at any other time during the year, but vertical objects still cast shadows. He then measured the length of the shadow vertical stick during the solstice noon, he could figure out what *angle* the Sun made with the vertical direction at Alexandria. In practical terms a curvature was detected with the difference in angle of the shadows and since the sun emits even and parallel rays, earth must be the source of the curvature which means that

Zak Ettamymy

it is curved, Eratosthenes went beyond this to measure the circumference of earth and the distance from earth to the moon using the same method and with an astonishing accuracy.

Humanity struggled to keep up with Eratosthenes' advanced methodology, his analysis was ahead of the time, so people went on believing in the comfortable yet erroneous notion that earth is flat, and the fact that earth was the center of the universe, also known as Geocentrism doctrine, which was championed by the religious people who used it to prove the importance of humans in the universe in harmony with Plato's model which claimed that earth is the center of the universe and all other stars and planets hover around it. This notion satisfied the religious leaders and those with close relationships with the dominant power of the time the priesthood and the religious establishments.

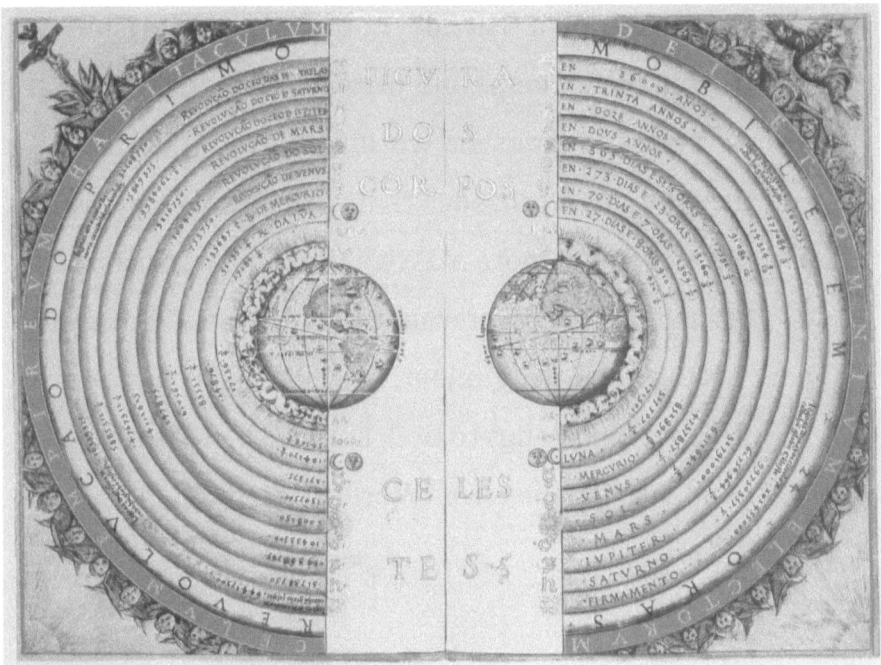

Figure of the heavenly bodies — *An illustration of the Ptolemaic geocentric system by Portuguese cosmographer and cartographer Bartolomeu Velho*

In the midst of an age of religious doctrines that forbid and challenged any scientific advances in this field there were people of knowledge who defied this model with scientific and sometimes religious proofs, such as: Fakhr al-Din al-Razi (1149–1209), in dealing with his conception of physics and the physical world in his *Motalib*, rejects the Aristotelian (Greek)

Zak Ettamymy

and Avicennian (Islamic) notion of geocentrisim
within the universe, but instead argues that there are
"a thousand thousand worlds (*alfa alfi 'awalim*)
beyond this world such that each one of those worlds
be bigger and more massive than this world as well
as having the like of what this world has." To support
his theological argument, he cites the Qur'anic verse,
"All praise belongs to God, Lord of the Worlds,"
emphasizing the term "Worlds."

400 years later Copernicus somewhat defended Al
Razi's point of view and switched the center of the
universe to the sun, although Al Razi's idea was

beyond discrediting the geocentrism and more in the realm of the multiverse but Copernicus was on the right track. After Copernicus came Galileo, who in 1610 published *Sidereus Nuncius* describing the findings of his observations with the telescope which he built and which supported Copernicus model. Astronomy become a hobby for all people at this period because telescopes became easier and easier to make or acquire. During this period there were many discoveries pertaining to the planets of the solar system, but no significant discoveries were made until 1687 with the publishing of the book *Philosophiae Naturalis Principia Mathematica By Issac Newton.*

The Islamic era of enlightenment and advances in science pushed the field of astronomy into a new stage of discoveries. Scientists of Baghdad, Cairo, Fes and Cordoba challenged the classic astronomy and prepared humanity to what was later called the renaissance. The Europeans didn't use the Greek methodology because it was a thousand years too old; the Islamic scholars were the ones who kept the enlightenment of humanity alive for a thousand years. To the credit of Neil de Grasse Tyson, he always tries to tie modern astronomy and science to an era that was always and purposely ignored by the western world, the time when Islamic scientists took on the burden to safe guard and advance this field. Neil always mentions their contributions, unfortunately he is one of the very few to do so.

Issac Newton

Newtonian Law of Gravity

Isaac Newton was not exempt from the problems that plagued the science of astrophysics; he was accused of plagiarism when he published *Principia* at the royal Society by a contender called Robert Hooke. Robert Hooke published his ideas about the "*System of the World*" in the 1660s, when he read at the Royal Society on March 21st, 1666 a paper "*On gravity*", "concerning the inflection of a direct motion into a curve by a supervening attractive principle", and he published them again in somewhat developed form in 1674, as an addition to "An Attempt to Prove the Motion of the Earth from Observations". Hooke announced in 1674 that he planned to "explain a System of the World differing in many particulars from any yet known", based on three "Suppositions": that "all Celestial Bodies whatsoever, have an attraction or gravitating power towards their own Centers" [and] "they do also attract all the other Celestial Bodies that are within the sphere of their activity"; that "all bodies whatsoever that are put into

Zak Ettamymy

a direct and simple motion, will so continue to move forward in a straight line, till they are by some other effectual powers deflected and bent..."; and that "these attractive powers are so much the more powerful in operating, by how much the nearer the body wrought upon is to their own Centers". Thus Hooke clearly postulated mutual attractions between the Sun and planets, in a way that increased with nearness to the attracting body, together with a principle of linear inertia.

Hooke's statements up to 1674 made no mention, however, that an inverse square law applies or might apply to these attractions. Hooke's gravitation was also not yet universal, though it approached universality more closely than previous hypotheses. He also did not provide accompanying evidence or mathematical demonstration. On the latter two aspects, Hooke himself stated in 1674: "Now what these several degrees [of attraction] are I have not yet experimentally verified"; and as to his whole proposal: "This I only hint at present", "having myself many other things in hand which I would first complete, and therefore cannot so well attend it" (i.e. "prosecuting this Inquiry"). It was later on, in writing on 6 January 1679|80 to Newton, that Hooke communicated his "supposition ... that the Attraction always is in a duplicate proportion to the Distance from the Center Reciprocal, and Consequently that the Velocity will be in a sub duplicate proportion to the Attraction and Consequently as Kepler Supposes

Reciprocal to the Distance. (The inference about the velocity was incorrect.)

Hooke's correspondence of 1679-1680 with Newton mentioned not only this inverse square supposition for the decline of attraction with increasing distance, but also, in Hooke's opening letter to Newton, of 24 November 1679, an approach of "compounding the celestial motions of the planets of a direct motion by the tangent & an attractive motion towards the central body Newton:

 On the other hand, Newton did accept and acknowledge, in all editions of the 'Principia', that Hooke (but not exclusively Hooke) had separately appreciated the inverse square law in the solar system. Newton acknowledged Wren, Hooke and Halley in this connection in the Scholium to Proposition 4 in Book 1. Newton also acknowledged to Halley that his correspondence with Hooke in 1679-80 had reawakened his dormant interest in astronomical matters, but that did not mean, according to Newton, that Hooke had told Newton anything new or original: "yet am I not beholden to him for any light into that business but only for the diversion he gave me from my other studies to think on these things & for his dogmatical in writing as if he had found the motion in the Ellipsis, which inclined me to try it"

Modern controversy

Since the time of Newton and Hooke, scholarly discussion has also touched on the question of whether Hooke's 1679 mention of 'compounding the motions' provided Newton with something new and valuable, even though that was not a claim actually voiced by Hooke at the time. As described above, Newton's manuscripts of the 1660s do show him actually combining tangential motion with the effects of radically directed force or Endeavour, for example in his derivation of the inverse square relation for the circular case. They also show Newton clearly expressing the concept of linear inertia—for which he was indebted to Descartes' work, published in 1644

Zak Ettamymy

(as Hooke probably was). These matters do not appear to have been learned by Newton from Hooke.

Nevertheless, a number of authors have had more to say about what Newton gained from Hooke and some aspects remain controversial. The fact that most of Hooke's private papers had been destroyed or have disappeared does not help to establish the truth.

Newton's role in relation to the inverse square law was not as it has sometimes been represented. He did not claim to think it up as a bare idea. What Newton did was to show how the inverse-square law of attraction had many necessary mathematical connections with observable features of the motions of bodies in the solar system; and that they were related in such a way that the observational evidence and the mathematical demonstrations, taken together, gave reason to believe that the inverse square law was not just approximately true but exactly true (to the accuracy achievable in Newton's time and for about two centuries afterwards – and with some loose ends of points that could not yet be certainly examined, where the implications of the theory had not yet been adequately identified or calculated).

About thirty years after Newton's death in 1727, Alexis Clairaut, a mathematical astronomer eminent in his own right in the field of gravitational studies, wrote after reviewing what Hooke published, that "One must not think that this idea ... of Hooke diminishes Newton's glory"; and that "the example of Hooke" serves "to show what a distance there is between a truth that is glimpsed and a truth that is demonstrated. (Wikipedia and Encyclopedia Britannica)

Zak Ettamymy

The Big Bang Theory (BBT)

Since Newton's apple story, people thought of the universe as a place where massive structures shaped and ruled the space which at the time only consisted of the Milky Way Galaxy. The beginning of the universe was never an important topic because the universe was thought of as static, it was thought to have always existed and looked the way it did and was made to look and behave in the way it looked and behaved for eternity, so searching for a beginning was like searching for the edge of the flat earth back in the bronze age, a futile exercise and an unnecessary endeavor. Even great minds like Albert Einstein failed to see that the universe was not static and that it was not inherently ageless. However his General Relativity theory was a revolution in the field of astronomy because it introduced time and space in the gravity mechanics, his belief in the good old universe crippled the theory at its infancy. His theory meant that massive objects not only distorted and pulled other masses as Newton said two centuries earlier, Einstein proved beyond doubt that space and time curve in the presence of such mass, he almost accomplished a perfect formula except for his stubborn belief in a static universe, he later solved his uncompleted formula with a cosmological constant. This value became irrelevant later because Hubble and Le Maitre proved the universe to be expanding and the band aid fix that Einstein used to keep his universe static was discarded.

Right after Einstein 's discovery the world scientists, Europeans, Russians and mainly Americans went to work, to produce results from Einstein's equation, Hubble was at the right place and at the right time, his 100-inch telescope over Mount-Wilson gave him a winning edge, he observed what is now known to be Nebulae from different galaxies and compared them to known stars, he discovered that there were flying away from each other, this meant one thing the universe was expanding and the expansion meant that at a time T-0 the universe was in one place. The theory predicts that if you rewind the video tape, the universe would shrink into one small place. However Hubble never went as far as calling the discovery the Big Bang. English astronomer Fred Hoyle is credited the term "Big Bang" during a 1949 BBC radio program. It is popularly reported that Hoyle, who favored an alternative "steady state" model, intended this to be pejorative but it caught on. ,the scientific community stayed divided between the steady state theory and the big bang until 1964 when two American scientists Penzias and Wilson stumbled on the microwave background radiation that permeates the universe, discovered by coincidence the radiations which lead the way to support once for all the Big Bang Theory.

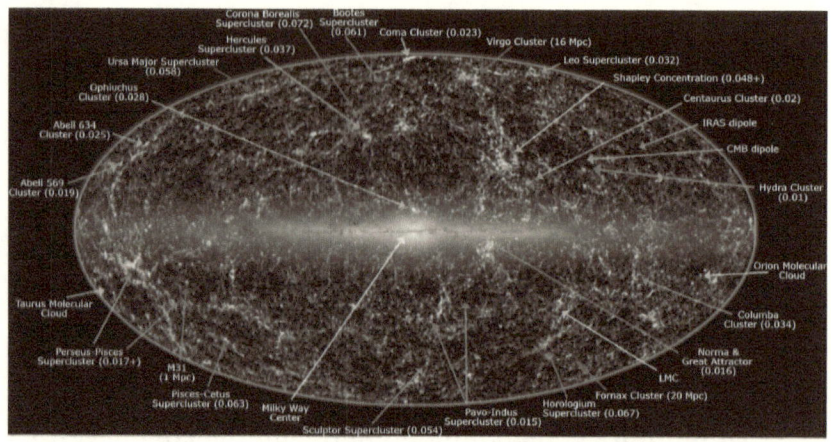

The Big Bang Theory has answered many centuries-
old questions but it added more unanswered
questions: the Big Bang is a beginning of time, space
and matter, the problem here is, closer we get to this
beginning harder it is to keep our physics in check,
the theory says that all matter in the universe and all
space and time and energy were all compressed into
an infinitely small mass with an infinite density and
infinite temperature and that we can only get as close
to it as a plank time = 5.39 x10-44 second (a plank
time is the time a plank unit travels at speed of light a
cross a plank length which is 1.616 199 (97) x10-35,
as the matter fact there are more plank time in a

Zak Ettamymy

second than seconds since the binging of the universe after this time all known laws of physics break down faced with the ultimate singularity which is another way of saying: *the power of infinity on Laws of physics or in laymen's terms on reality.* The theory supporters (at this point all scientists do support the BBT) argue that the only way for the universe to have a beginning and the only way to explain the expansion of the universe is with the BBT, a birth had to be introduced to make sense of mankind's legacy because having a thinking species contemplating the laws of the universe but failing to explain the birth of such universe was seen as a colossal failure, after all humans set out to prove the meaning and origin of the universe for religious and scientific reasons since intelligence become a tool.

But the BBT adds more confusion and may have had counterproductive results on the scientific quest, it seems that scientists rushed to this theory without knowing that when asked about what was there before the Big Bang they would have nothing to say,

Zak Ettamymy

and even more damaging no discoveries in this field are in the works, the Hedron Colliders can never tell us what was there before the Big Bang.

The concerning aspect of this paradox and confusion is that it is taught to the children in schools and on TV without a disclaimer or clarification that specifically indicates that it's a theory and as many other theories they were proven to be a mere step towards more meaningful results or sometimes even just a waste of time, the fact that we talk about a theory of the beginning of the universe while ignoring all laws of physics promotes illogical thinking among the upcoming regeneration. Kids with a little bit of curiosity and a sense of logic would press you to tell them what was there before the bang, and since scientists always explain the Big Bang with an explosion, if we run the tape backward we can demonstrate that everything was compressed into a singular place and space , this is erroneous because if we run the tape backward we would take the sun and add all matter of our galaxy and shove it into a

Zak Ettamymy

smaller and smaller space, gravity that is holding the sun intact will breakdown and we'll end up with what's called a black hole. Then how can the universe be so small if even much bigger masses can't be compressed more than a certain level? Meaning if we run the video tape back we will not see a smooth compressed universe into a one singular mass, we will see billions of failed universes if the black hole theory holds.

The question of what was before the Big Bang is usually handled by the astrophysicists with a philosophical approach intended to confuse or at least distract the audience than to face the reality and admit that they have absolutely no idea. Their answers are usually hints to possibilities not answers which none make sense in the logical sense. But yet we seem to scrutinize our doctors, politicians, teachers but not the astronomers. Why? Maybe because they're the only people with a knowledge that is beyond our realm of expertise and because the Big Bang remains a tense subject that scientists keep clear of as much as they can, they like to talk about it in a more distant memory that almost has no bearing on the today's astronomy.

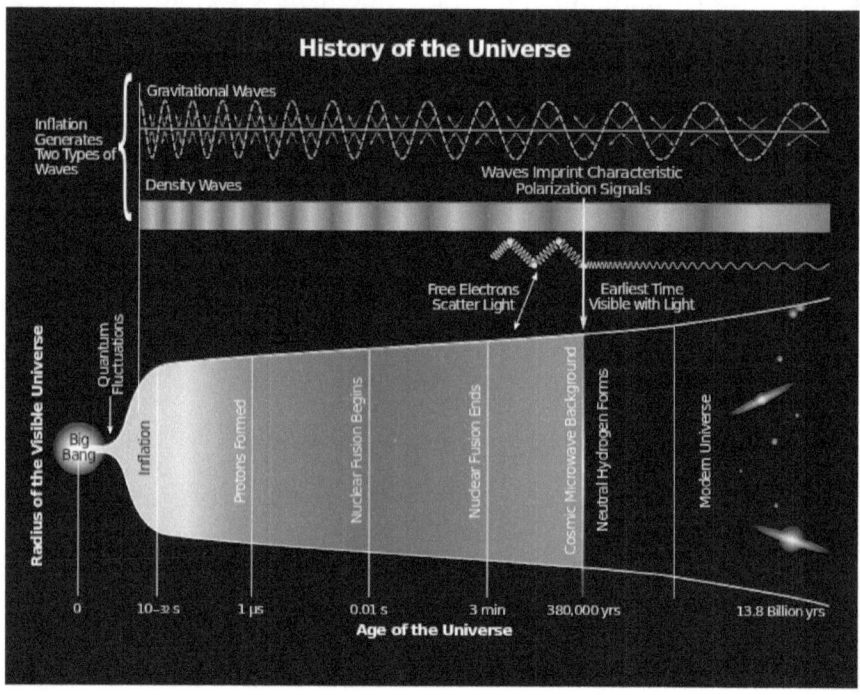

BBT

Another problem with the theory is the fact that it relies on Einstein's theory of relativity or $E=MC^2$, this formula has proven to be correct and true in a sense that a tremendous burst of energy can be produced

Zak Ettamymy

and unleashed by changing the property of a mass, the atomic bonding of an atom has a great energy and the formula was proven to be correct, unfortunately, with the invention of the atom bomb. The Big Bang supporters however use the reverse of the formula to prove its validity, meaning the energy this time divided by the constant can produce mass. In clean mathematics this is 100% correct however in a plank time *(0.000000000000000000000000000000000000539 seconds)* this formula doesn't not hold its water especially when we're faced with the infinity of energy and infinity of temperature $E=MC^2$ can't be used because at the event-horizon Laws of physics don't work anymore. Furthermore the astrophysicists and astronomers talk about the Big Bang as being this unprovoked explosion, in a random real of no time and no space and no universe and while the explosion was underway there was only energy in the first plank times, no particles existed yet, but this energy created the mass that is here now as the universe cooled gradually, but the question remains: Why don't we see this happen right now? since the universe is expanding which means that we are somehow still experiencing the effect of the explosion (BB), why can't we see an energy give birth to matter, in other words why isn't matter being produced today, new matter, we know that there is not a lack of energy in the universe there are phenomenon that are super energetic in the universe right now but until this day no one could point at a place in the universe where energy is being converted into matter. Some may say Nebulae are the nursery of the starts, that's where an example of how matter is being fabricated, but Nebulae work with the force of gravity to compress already existing matter into stars which is different from $M=E/C^2$. Since the Big Bang or the beginning of the universe there has not been any new matter produced with this formula, matter is recycled compressed, changed nature or shape but never made from energy. Nothing is made in the universe everything has been here since the beginning of time, whether that is the BB or some

Zak Ettamymy

other beginning. For instance a star burns Hydrogen into Helium and when it heats up even more it starts to change Helium into other more complex elements such as Iron and Carbon but these elements were always in the universe, only their electrons get snatched away or added through the process of fusion but the quantity of electrons in the universe has been and is and will always be the same number unchanged since T0. So the fact that energy has created everything in the universe is questionable to put it mildly.

My theory is that the universe has always existed in many shapes and forms and the elements that are here today have always been part of it, yet the energy of the Big Bang perhaps just bonded these elements into atoms meaning all types of forces: atomic, electromagnetic and gravity all worked together to assemble the universe into small entities (quantum and bosoms) and into bigger structure galaxies, stars and planets. But nothing making something is an insane proposition and shoving it down people's throat is not a way to prove its validity.

The Universe or the World

The World was the Universe until the middle ages, when Man started to open the horizons and started to interact with the worlds around him. A World is now a philosophical concept of everything that involves the beings, the Universe is more of a space that contains matter and energy. The two have been diverging since Galilee took a first look at his telescope... the Universe became a cold dark unknown place and the World become a smaller unity that is warm and friendly, our World included earth only but now the moon and perhaps Mars could be called our world because we're in a constant contact with them, it s our warmth that separates a World from a Universe. The World lost its scientific meaning because it failed to mean "everything" since the "everything "has been expanding to more and more territories, the World of pre-1492 didn't include the Americas and the World of Alexander the Great only included the Mediterranean sea and Persia, India and China. The World changed its meaning because it changed its boundaries as people sailed across oceans and journeyed across the skies to the moon and sent man made robots to Mars and other far planets. But what's intriguing about the World is that now it could also be related to a certain species especially with the rise of science fiction in pop culture. Aliens also have their Worlds so the World became a notion of consciousness limited area, in the mind of a science fiction writer there is alien World and there is our world all swimming in a sea called the Universe. Another way of looking at this in a simplistic way is whenever there is nothing or no intelligent life it's a Universe and whenever there is consciousness or intelligent life it's a World. The reason why we experience a confusing relationship between World and Universe is because in reality we still only know of one world and the day we find an intelligent life somewhere in the Universe there will be a second World that will clarify the confusion once for all. At that point there will be a Universe separating two Worlds.

Zak Ettamymy

Zak Ettamymy

The Universe or the Multiverse

The universe can be used to calculate its approximate age by extrapolating backwards in time, almost like using the rings on a tree to know its age and same thing with the denture or skeleton of an animal, our universe is said to be 13.7 billion-years old, based on a careful regression of events which is the widely accepted version. At this age the universe seems very young if compared to the opposite theory which claims that the universe had always existed. The notion of a universe as unique came to humans by nature. The world/universe had to be unique and any deviation from the fact that we, our earth and our universe are all unique was seen as a sin or anti religion. In reality it was anti establishment: the church the kings and the priests to claim that Men, earth and the small world around them are not the center and the reason of the universe. Blasphemy was a sin punishable by death, the religious establishments killed scientists for daring to shift the center of the universe from us, our planet and our universe… A thousand years ago Al razi played around this notion and with courage he used the Quran to explain that there are "worlds" out there, in an old sense of the word" Multiverse".

In 1961, Robert Dicke noted that the age of the universe, as seen by living observers, cannot be random. Instead, biological factors constrain the universe to be more or less in a "golden age," neither too young nor too old. If the universe were one tenth as old as its present age, there would not have been sufficient time to build up life and appreciable levels of metallicity (levels of element besides hydrogen and helium) especially carbon, by nucleosynthesis, which are the building blocks of life. But in reality the universe, if it is 13.7 billion-years old, it's an infant especially if we consider the fact that a simple star like our sun can live to be 10 to 12 billion years old, scientists predict the end of the universe to be in a big crunch or deep freeze in over a trillion years from now, if this is the case the universe is a few hours old in human life's perspective. Calling the universe age the right time to sustain life is like saying the same thing about a baby who was just born. The universe is too young to consider this stage as the golden age... there will be many more events in the universe that will shape many more outcomes. And we may not be here to witness it. In 1584, the Italian philosopher and astronomer Giordano Bruno introduced an unbounded universe in *On the Infinite Universe and Worlds*: "Innumerable suns exist; innumerable earths revolve around these suns in a manner similar to the way the seven planets revolve around our sun. Living beings inhabit these worlds."

Zak Ettamymy

How did modern scientists think of the Mulitiverse? 800 years after Al Razi and 300 years after Bruno no one really paid attention to the possibility of more than this universe, humans thought of the Milky Way galaxy as the entire universe. The end of 19th century and the beginning of the 20th century saw a jump in scientific daring after the establishment of many theories that led to big ideas (Relativity Theory, Blackholes, Big Bang Radiation, Quantum Physics etc..) scientists dared to ask the big questions: Are we alone? Is this it?

The result was a fury of theories all led to two polarizations: multi-universe supporters and universe supporters just as it was the case for the flat earth supporters and round earth supporters a few centuries earlier.

The most notable scientists right now support the possibility of multiverse based on:

1-Quantum Physics: Quantum Physics led the way to a new look at the big picture of the universe from the small universe prospective. Quantum physics predict infinity of possibilities of any existence, all co-existing at the same time and the same space. Many universes could occupy the same space but in different properties, matter, space and time are all elements of an equation which results to infinity of existences.

2-Holographic Principal: Closely correlated with the String theory but with a little difference in how the information makes matter or the universe or many universes.

3-Blackhole Theory: Based on this theory Black holes are portals to different universes, everything that goes through would show up in a completely different universe, some even think that the Big Bang is an explosion of one black hole from older universe to ours.

4-String theory: this is a theory that adds more dimensions to demonstrate existence of more universes.

Zak Ettamymy

The Concept of Infinity ∞

At early age we learn that numbers have no limit, you can keep counting numbers forever, all you have to do is keep adding a One and you'll find yourself with a bigger number than your previous biggest number each time and so on. But this infinity in abstract and does not bother our outlook on the universe because we never really deal with big numbers in our lives, no one has billions and trillion of anything, we don't think about handling or dealing with astronomical numbers.

Google is perhaps the most known number in today's computerized existence, it means 10 to the power of 100, and it was invented by a nine year old boy. The name became a household especially after the internet giant called its search engine many other products google. There is also Google plex which is 10 to the power of Google. This number is larger than all the atoms in the universe; it is even larger than all the plank times since the Big Bang. However as large these numbers may be, infinity is much larger. But what does infinity really mean?

Well infinity is a valid concept in math but its physical presence or reality is impossible to detect, on the small or big scale, infinitely small items and infinitely big items are equally ambiguous: For example, if you divide a number by 2 and then the divide the result by 2 you' ll spend your life time dividing these results and you'll never be able to find a finite solution, there is also π, you can keep writing the decimals forever and without a patterned or repeated sequences to infinity. If you try to find the square root of 2 you'll also find yourself in the realm of infinity.

We work with infinity in Math without too much fuss about what it really means, I remember when I was in middle school and having to deal with results that lead to infinity like: $\lim_{x \to \infty} (f(x)^2) = \infty$ it makes sense that as large a number can be as even larger its square would be, but in real life nothing should lead to infinity not the space nor time because they re bounded by the laws of physics. It is easy to write Google plex on a piece of paper as a formula or a symbol, but if you set out to write all the zeros of Google plex, the observable universe will not be sufficient to fit them all and if we use the Google plex as a time measurement, in one Google plex you can fit billions generations of universes each with trillions of years life span, And by the way Google plex plus 1 is even larger than Google plex… This magnitude of space and time of a finite number Google plex should be a deterrent for us to consider infinity as unit of measurement or a reality.

Infinite universe is a universe that makes no sense in reality, many scientists and physicists insist on Alan's Guth's theory of inflation, I find it to be nothing but a quick-fix formula because it provides us with a reason to continue believing in the Big Bang Theory since without it the Big Bang would always be missing a link to infinity. Alan Guth proposed the theory of inflation to duct tape the Big Bang Theory, he suggests that the universe experienced a brief period of inflation or a burst, almost an explosion within the explosion except this time the second explosion defied the laws of physics. How? : Based on his theory, during the inflation the universe expanded with a rate faster than speed of light. The infinity epoch started at 10-60 sec after the Big Bang and ended at 10-32 sec later. This theory has seen its share of hypes and setbacks, as recently as Sep 2014 there was a substantial reason to doubt it because since its introduction, this theory has not been measured or observed. The theory claims that it's possible for the universe to expand faster than light because in reality matter stays in its place only empty space between matters is what expands, which is a way to get around the problem with having particles travel at a speed beyond that of the speed of light. Yet even with the unbelievable notion that space could just burst into existence between two or more particles of matter, the question remains: was not that a push of these particle? No matter how you see it, the particles will get pushed at a speed faster that of light, which is a big No No in physics. Any body that is expanding can never be infinite.

The Realm of Consciousness

سورة الحجرات

بِسْمِ اللَّهِ الرَّحْمَٰنِ الرَّحِيمِ

يَا أَيُّهَا الَّذِينَ آمَنُوا لَا تُقَدِّمُوا بَيْنَ يَدَيِ اللَّهِ وَرَسُولِهِ وَاتَّقُوا اللَّهَ إِنَّ اللَّهَ سَمِيعٌ عَلِيمٌ ۝ يَا أَيُّهَا الَّذِينَ آمَنُوا لَا تَرْفَعُوا أَصْوَاتَكُمْ فَوْقَ صَوْتِ النَّبِيِّ وَلَا تَجْهَرُوا لَهُ بِالْقَوْلِ كَجَهْرِ بَعْضِكُمْ لِبَعْضٍ أَن تَحْبَطَ أَعْمَالُكُمْ وَأَنتُمْ لَا تَشْعُرُونَ ۝ إِنَّ الَّذِينَ يَغُضُّونَ أَصْوَاتَهُمْ عِندَ رَسُولِ اللَّهِ أُولَٰئِكَ الَّذِينَ امْتَحَنَ اللَّهُ قُلُوبَهُمْ لِلتَّقْوَىٰ لَهُم مَّغْفِرَةٌ وَأَجْرٌ عَظِيمٌ ۝ إِنَّ الَّذِينَ يُنَادُونَكَ مِن وَرَاءِ الْحُجُرَاتِ أَكْثَرُهُمْ لَا يَعْقِلُونَ ۝ وَلَوْ أَنَّهُمْ صَبَرُوا حَتَّىٰ تَخْرُجَ إِلَيْهِمْ لَكَانَ خَيْرًا لَّهُمْ وَاللَّهُ غَفُورٌ رَّحِيمٌ ۝ يَا أَيُّهَا الَّذِينَ آمَنُوا إِن جَاءَكُمْ فَاسِقٌ بِنَبَإٍ فَتَبَيَّنُوا أَن تُصِيبُوا قَوْمًا بِجَهَالَةٍ فَتُصْبِحُوا عَلَىٰ مَا فَعَلْتُمْ نَادِمِينَ ۝ وَاعْلَمُوا أَنَّ فِيكُمْ رَسُولَ اللَّهِ لَوْ يُطِيعُكُمْ فِي كَثِيرٍ مِّنَ الْأَمْرِ لَعَنِتُّمْ وَلَٰكِنَّ اللَّهَ حَبَّبَ إِلَيْكُمُ الْإِيمَانَ وَزَيَّنَهُ فِي قُلُوبِكُمْ وَكَرَّهَ إِلَيْكُمُ الْكُفْرَ وَالْفُسُوقَ وَالْعِصْيَانَ أُولَٰئِكَ هُمُ الرَّاشِدُونَ ۝ فَضْلًا مِّنَ اللَّهِ وَنِعْمَةً وَاللَّهُ عَلِيمٌ حَكِيمٌ ۝ وَإِن طَائِفَتَانِ مِنَ الْمُؤْمِنِينَ اقْتَتَلُوا فَأَصْلِحُوا بَيْنَهُمَا فَإِن بَغَتْ إِحْدَاهُمَا عَلَى الْأُخْرَىٰ فَقَاتِلُوا الَّتِي تَبْغِي حَتَّىٰ

Qur'ān verse 49:13 states: "O humankind! We have created you out of male and female and constituted you into different groups and societies, so that you may come to know each other-the noblest of you, in the sight of God, are the ones possessing taqwā."

Zak Ettamymy

In Islam, according to eminent theologians such as Al-Ghazali, although events are ordained (and written by God in al-Lawh al-Mahfūz, the Preserved Tablet), humans possess free will to choose between wrong and right, and are thus responsible for their actions; the conscience being a dynamic personal connection to God enhanced by knowledge and practice of the Five Pillars of Islam, deeds of piety, repentance, self-discipline and prayer; and disintegrated and metaphorically covered in blackness through sinful acts Marshall Hodgson wrote the three-volume work: The Venture of Islam: Conscience and History in a World Civilization.

In the Protestant Christian tradition, John Calvin saw conscience as a battleground: "[...] the enemies who rise up in our conscience against his Kingdom and hinder his decrees prove that God's throne is not firmly established therein". Many Christians regard following one's conscience as important as, or even more important than, obeying human authority.

Zak Ettamymy

What is Consciousness?

The difference between humans and all other beings is the ability and wiliness to analyze; think and take a conscience decision not a survival one, to question and try to make sense of the self. Our consciousness is so powerful it is the precursor of existence itself based on Quantum Physics. Free will is not exercised within the cells of the brain, it is the language of the soul of the person, the distinction between the physical brain functionality and the consciousness is the same distinction between the driver and the vehicle. The brain is a vehicle that consciousness uses to express its existence, to leave an imprint of its thoughts and to propagate its purpose.

Other definitions of consciousness are as descriptive: awareness, wakefulness and selfhood. They all mean one thing, the awareness of the self and the outside world. We feel and interact with our self in a different manner than we do with the rest. Expressing thoughts is a great indication of consciousness but they can also be expressed in different ways, such as emotions and feelings.

The peculiar thing about consciousness is that although it seems to be independent of the body it occupies, it never leaves the body to another one, you never feel that you are trapped inside someone else's body, your awareness is always comfortable with the body it occupies, you may or may not like the body you're in but you're in your right body. It seems that consciousness is assigned to one body and a body is assigned to one consciousness.

Other types of consciousness such as animal consciousness is difficult to talk about because there is a barrier between us "the observer" and them in expressing consciousness and although emotional awareness is considered a form of consciousness it is an immeasurable state. Until further discoveries the animal consciousness is considered a survival set of behaviors that give the impression of awareness.

There are states when consciousness could behave erratically or stop functioning all together such cases are rare but they're well documented. Example: Anosognosia, a striking disorder that makes a person unaware of the state he is in especially in cases of stroke, the patient could lose his sight for example but keeps claiming that he/she can still see. Other minor loss of consciousness is sleepwalking, a person's body become independent of the consciousness and behaves in an automated way.

In religion consciousness is the most important state of a person, because it is considered the driver of the body, all bodily sins are provoked by the consciousness and it s held responsible based on the fact that only it holds the burden of morality.

The medieval Muslim philosopher and physician Muhammad ibn Zakariya al-Razi believed

Zak Ettamymy

in a close relationship between conscience or spiritual integrity and physical health; rather than being self-indulgent, man should pursue knowledge, use his intellect and apply justice in his life. The medieval Islamic philosopher Avicenna, whilst imprisoned in the castle of Fardajan near Hamadhan, wrote his famous isolated-but-awake "Floating Man" sensory deprivation thought experiment to explore the ideas of human self-awareness and the substantiality of the soul; his hypothesis being that it is through intelligence, particularly the active intellect, that God communicates truth to the human mind or conscience. According to the Islamic Sufis conscience allows Allah to guide people to the marifa, the peace or "light upon light" experienced where a Muslim's prayers lead to a melting away of the self in the inner knowledge of God; this foreshadowing the eternal Paradise depicted in the Quran

Zak Ettamymy

Quantum Physics and the Entanglement

Quantum is the minimum amount of any physical entity in motion or in direct interaction with others. Example of Quantum is the photon of light. Consequently Quantum Physics is the science that deals with the mechanics of the small particles in the sub atomic scale.

As almost all sciences, Quantum Physics answered many questions but added more unsolved mysteries to the quest of a unified theory of everything, the illogical behavior of matter at the Quantum level beaks all laws of physics as does the Big Bang and infinity on a larger scale.

In a nutshell Quantum Mechanics is based on two things: A formula known as the uncertainly principal $\sigma_x \sigma_p \geq \dfrac{\hbar}{2}$ and the phenomenon of entanglement.

The Uncertainty Principle: As mentioned before humans being part of the experiment find it hard to judge the experiment, meaning the Big Bang, gravity etc..All are phenomenon that affect the observer and distort his/her ability to play the role of unbiased observer. In Quantum Physics the observer effect is magnified to an unbelievable magnitude. Quantum Mechanics' experiments were done in early 20th century mainly by Werner Heisenberg at first and it took a few years for the results to be fully adhered and accepted, especially since Albert Einstein rebuffed their validity at first.

In the experiments, the scientists shot one electron (a small mass unit) into a screen and as predicted the particle arrived and was observed as such. The experiment then continued now with two electrons and again two particles were found on the screen. The third experiment was done without observation, meaning the particle was shot without observing the process, the result of this third experiment was drastically different than that with an observer, the one particle that left the source actually made multiple arriving points on the screen, this meant

Zak Ettamymy

that we started with one particle being emitted from the particle source but ended up being multiple particles on the arrival. This perplexed the scientists, they then conducted a fourth experiment this time they shot the particle and didn't not observe at first but in the middle of the experiment they initiated the observation, the result was even more perplexing, the particle went from multiple particles to one as soon as the observation was introduced and a fifth experiment introduced a delayed observation and the particle behaved in an even more bizarre way, it went back in time and changed its property into one particle and got projected on the screen as such.

These five experiments lead to the uncertainty principal which in essence demonstrates that a particle is what it is perceived to be only with an observer, otherwise a particle is a wave of possibilities when the observer is not aware of the experiment, this principal challenges even the best minds, Einstein once said "I can't not accept that when I am not looking the Moon is not there" but the

Zak Ettamymy

experiment is clear: *on a small scale a particle is there because we're observing it otherwise it is everywhere.*

The implications of the uncertainty principal: Out of thousands of laws and principals, I can't find one that introduced so much chaos into the way the universe is perceived to function. The uncertainty principal basically throws all Newtonian Mechanics out the window and it challenges Einstein's theory of Relativity. Not only that it makes the universe, life, consciousness and everything else in a soup of endless possibilities where you are all that you can be and more...Furthermore that everything in the universe is based on consciousness, because the observer's consciousness is what freezes everything into being, meaning the Sun is the Sun because of us observing it, consequently we are making the universe by just consciously being aware of its existence. So the big idea behind the uncertainty principal is that existence is the product of consciousness and when it is absent the universe swims in a sea of strange and parallel existences.

Zak Ettamymy

So what really exists in the universe is our consciousness, the rest (that is everything in the universe including our bodies) is all made up. This is where our astrophysicists failed to connect the dots, the Big Bang is perhaps not an explosion of matter, it's the birth of energy that produced consciousness. It doesn't not need to be attached to a human body to function it could have made the universe based on the uncertainly principal, and this is one of many possibilities that could solve the divergence between religion and science.

Entanglement: Another aspect of Quantum physics that chocks scientists and average people alike, the theory explains that if we take two or more objects with same Quantum slate references, their behavior, in this case let's call it spinning since everything in the universe spins, all objects will spin in a way that is in full correlation with the other object regardless of the distance.

Example: if we take object A and put it 100 million light years away from object B , if they're in full Quantum Slate if object A turns clock wise, object B will spin in the opposite direction instantly. This may not be a farfetched idea but consider the distance between them 100 million light years meaning if information of the state on one object at the speed of light will reach the other object 100 Million years later. So the immediacy of their correlation is done faster than the speed of light which means the two objects communicate in a way that is physically impossible...but they do.

The implications of such theory are endless; you can imagine that in a far away universe there is you in full correlation with you and your actions are in reality just coded with immediate instructions. This could also be explained with the notion of destiny v.s choice

Zak Ettamymy

Entropy and the Arrow of Time

We know what time it is, we can waste time and we can think of old times but it seems that the direction of time is a constant present-to-future progression and there is almost no way for humans to change this direction, stop or expedite events. Time runs independently of us and since we're the only entity that is watching the progression of time (as far as we can see and or experience) we are holding the watch of this universe and all events in it are based on our concept of time. But based on the relativity theory, time as well as matter and space can get influenced and distorted by gravity, so time can be stopped or slowed down but even with the presence of very heavy objects that could distort the notion of time the direction is always constant: present to future. This means that even black holes can't reverse the progression of time, at the event horizon time almost freezes at a constant "NOW" but it never goes back even during the Big Bang the burst of matter and energy and time continued and never at any time, it somehow stopped and went back. This unwavering direction of time is caused by Entropy meaning the universe is in a constant progression from order to disorder based on the second law of thermodynamics.

Entropy is still seen as one of the most misunderstood principal because it reflects on a bigger picture of time, space and matter. Entropy is a principal that explains equilibrium and the dissipation of energy and temperature but in a more simplistic argument it also explains the direction of time and the increase of chaos in the universe. Based on the fact that the universe was at perfect order at T0 or the Big Bang initiation, every second after that the universe has experienced a constant progression towards disorder. For example: if you accidently drop a glass of water from your hand and see it shatter on the floor, the state of the glass as a system went from order to disorder and this is only possible in the direction of order to disorder which causes time to have one direction, irreversible. You will never

Zak Ettamymy

experience the glass going from broken pieces to being a whole glass in your hand. Even traveling in time will not give you the ability to see things in reverse, theoretically you can go back in time and experience the past again from order to disorder meaning you'd see the glass as a whole and you'd see it dropped. But Entropy categorically forbids experiencing the reversal of order to disorder direction except in Quantum realm as explained previously. In the Quantum world many laws beak down to allow matter to go back in time and experience disorder to order direction

So based on the Entropy principal, time is trapped in the direction that entropy has set for it which is order to disorder, unlike other dimensions time cannot experience the same position twice, in space you can pass by point A more than one time but in time you cannot experience 3:10PM Dec 9th 2014 more than once. Time is a one way ticket from now to the future. The arrow of time was explained by Eddington as such:

Zak Ettamymy

1. It is vividly recognized by consciousness.
2. It is equally insisted on by our reasoning faculty, which tells us that a reversal of the arrow would render the external world nonsensical.
3. It makes no appearance in physical science except in the study of organization of a number of individuals.

Arrow of time can also be explained by the fact that cause precedes effects, everything in the universe had a cause, you'll never experience an effect before its cause otherwise the universe will be in chaos, otherwise things will happen independently of the things that cause them, you'll be born before your parents and you'd be older then get younger etc...

But the most disturbing repercussions of the arrow of time is the "now" the interesting thing about time is that we only own the present, the future is unknown but it s coming the past is well known but inaccessible, so we're left with the present but this present is so illusive we don't even have the "now".

Zak Ettamymy

Now of a second ago is not the now of now and as I try to contemplate the now it passes by me to become a past because we cannot stop time and call a specific moment a now we never really experience the present because as soon as we think about the present it already became a past and this illusiveness pushes us to live in a limbo between the past and the future. The present only exists on paper.

The arrow of time as complicated as it seems get more complicated with the speed of light. Light photons travel at a constant speed, this speed is the highest speed any object can reach. Since S=d/t time is deeply imbedded in the fabric of space and the limits of the speed in which an object can reach affects the time directly as such.

Let's consider two kids of the same age one sitting in planet earth the other biking in the opposite direction on a distant planet (100 Million light years away). Their present becomes drastically different when the biker on the distant planet stops biking and when we assess his time vs. that of the earth's kid we're in for a huge surprise.

At the moment the alien kid stops biking his present has become a distant future for the earth kid, the present of the alien kids is tens of years in the future for the earth time, this means although the alien kid was moving at a low speed he created an angle in time/space fabric although so small it got magnified due to the huge distance between the two kids and projected on earth as a future event. Think of the time space fabric as a loaf of bread, because of the biking movement the now which is represented here with a knife cutting the loaf of bread will cut the loaf in an angle which lands on earth future.

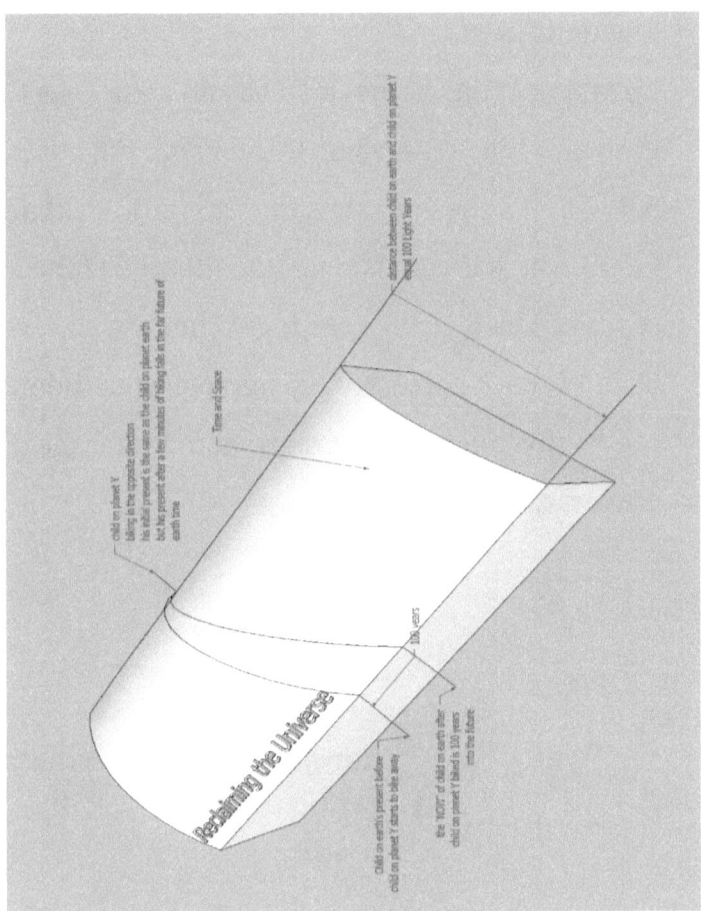

Fabric of time of space/time

The units of time:

Instant is a time zero which has no value, the first time unit that has a value is the plank time unit at 5.39 10-44s the largest is the **cosmological decade (CÐ)** which is a division of the lifetime of the cosmos. The divisions are logarithmic in size, with base 10. Each successive cosmological decade represents a ten-fold increase in the total age of the universe.

Units of time

Unit	Length, Duration and Size	Notes
instant	varies	loosely speaking, zero time (colloquially the term may be used in other ways)

Zak Ettamymy

Planck time unit	5.39×10^{-44} s	The duration light takes to travel one Planck length. Theorized to be the smallest duration measurement that will ever be possible, roughly 10^{-43} seconds.
yoctosecond	10^{-24} s	
jiffy	varies	in quantum physics, the duration light takes to travel one fermi (10^{-15}m, about the size of a nucleon) in a vacuum: about 3×10^{-24}s. In electronics, the duration for one alternating current power cycle (1/60 or 1/50 of a second).

		Also, an informal term for any unspecified short duration.
zeptosecond	10^{-21} s	
attosecond	10^{-18} s	shortest duration now measurable
femtosecond	10^{-15} s	pulse duration on fastest lasers
picosecond	10^{-12} s	
nanosecond	10^{-9} s	duration for molecules to fluoresce
shake	10^{-8} s	10 nanoseconds. Also a casual term for a short duration.

microsecond	10^{-6} s	
millisecond	0.001 s	shortest duration unit used on stopwatches
centisecond	0.01 s	used on some stopwatches
decisecond	0.1 s	used on some stopwatches
jiffy (electronics)	~1/50s to 1/60s	Used to measure the duration between alternating power cycles. Also a casual term for a short duration
second	1 sec	SI base unit

petasecond	1 quadrillion seconds	About 31.7 million years
era	varies	on the geological timescale, several hundred millions of years[42]
galactic year	Approximately 230 million years[43]	The duration it takes the Solar System to orbit the center of the Milky Way Galaxy one time.
eon	varies	on the geological timescale, half a billion years or more.[42] Also "an indefinite and very long period of time".[44]

gigaannum	1,000,000,000 years	1 billion years
exasecond	1 quintillion seconds	roughly 31.7 billion years, more than twice the age of the universe (on current estimates)
zettasecond	1 sextillion seconds	About 31.7 trillion years
yottasecond	1 septillion seconds	About 31.7 quadrillion years
cosmological decade	varies	

The Inconsistency of the Rhetoric

The field of Astronomy changed tremendously during the past 35 years, since Bob Sagan took to the airwaves and started teaching average American and others how to look up to the heavens and make sense of it all, things changed to an unrecognizable stage. In his TV series "Cosmos a Personal Voyage" Bob Sagan called the stars Suns, his TV program was a first encounter with the universe as a reality for many Americans, the invading Martians or other mid-20th century fictions were ignored and only science was discussed but he also satirized the fine line between science fiction and astronomy. Now we have astrophysicists and scientists talking about black holes, black energy and dark matter, none of which has been detected or seen, Astronomy has become bed time story directed at people who became passive consumers, overwhelmed by the strange names of stars and galaxies and the vast array of phenomenon and the complexity of the their universe, they gave up the most important element in science: the right to question, argue and put to test the theories. How can we explain the infinity as a realistic phenomenon or the fact that out of nothing the Big Bang burst out? How can we explain the expanding universe while they tell us that Andromeda is on a head-on collision with the Milky Way? How come we don't experience the expansion of the universe on earth? Why is the universe 13.7 billion-years old while it is more than 46 billion years across? Why is the Milky Way only a billion years younger than the universe? Woody Allen's character as kid being worried that the universe is expanding, and his mother assuring him that if the universe is expanding, at least Brooklyn is not and won't be for centuries to come. Astrophysicists made a name for themselves by telling us fascinating stories of worlds beyond our imagination, people like Steven Hawking, Neil De Grasse and Marc Greene, to mention a few, and others all captured our attention, imagination and innocence for decades by telling us how old the universe is, how spectacular the beginning of the world when it infolded and how the expansion of the universe shaped the fabric of the space we live in and so on.. Yet they failed to provide realistic and mathematical explanations of these phenomenon and when they're pressed for answered they evoke illogical explanations such as event horizon and the breakdown of physics and black this black that, we were told to accept the notion that this science is beyond our grasp and we should just listen and enjoy the voyage. But what these astrophysicists failed to notice is that they now exercise the same moratorium that the church subjected scientists to in the dark ages. In the 21st century a strong scientific society hijacked the discussion about the most important aspect of our existence, the universe, its origin and our place in it.

Astronomy has become a science of projects, and a showcase of machinery rather than a science of discovery, I find it interesting that astronomers go on TV to showcase big machines like the Hadron Collider in Cern, Switzerland and the technological marvels of these miles-long machines, they talk about them for hours rather than the actual experiments that these machines were constructed to conduct, throughout the documentaries they talk about the project of finding the Higgins Bosom or re-creating the conditions of the Big Bang or even finding extra terrestrial intelligence, all these machines become the focus of the discussion, while scientists completely brush off the fact that they never achieved results with these billion-Dollar gizmos, they never created the Big Bang conditions and they never answered if there is intelligent life out there. The age of discoveries has gone, now astronomers don't need to discover things, people like Pastor, Marie Currie or Graham Bell worked hard to advance human knowledge to a level astronomers have yet to achieve, the new astronomers are super stars who get the gratitude and appreciation for just showing up. Even the nostalgic notion of the lonely nights at the observatory on top of an isolated mountain has gone, they now have computers and camera scanning and taking pictures of the skies and feeding them as emails with data analysis which they can read on their tablets while having dinner with their families in the comfort of their living rooms. We are now in a stage of Hollywood, science museums and the planetariums across the world have teamed up to create a dramatization of astronomy and produced fairy tales and bed time stories for all us to feel cozy and fuzzy when we listen to Neil Degrasse, Mark Greene, Lawrence Krauss and Michio Kaku and Stephen Hawking telling us the story of us in a funny, easy and comforting delivery only rivaled by the politicians, all are complicit in the glazing of the universe and the intended consequences of separating the average person from his/her universe, they have been feeding us the nonsense of the 20th century discoveries produced by Hollywood with very friendly graphics that serve as a place holder to absent real results of real experiments. Tesla never had the audience that TV and Hollywood is now offering the new astronomers, he was shy and lived far less glamorous life but his inventions are real and are being used throughout the world on a daily basis.

On Monday Dec 8th 2014 NASA put out a news brief about a key evidence for a lake, which is consistent with the idea that Mars sustained some sort of life, most likely such thing is true NASA made this claim based on the Rover Curiosity discovery of a crater and the misleading part of this information is the explanation that came along the discovery. NASA claimed that "...period of time roughly 3.5 billion years ago when the crater was filled with water" so up to 3.5 billion years ago Mars had water already, So Mars cooled down and finished its formation in a billion year only. A rock planet of the size of Mars would need more than 2 Billion years to cool down for any liquid water to form on it. This type of misleading information is what makes average people shrug all these discoveries and almost look at the astronomers with a mocking air. This brings us to the dilemma of finding out the age of a star of a planet, if the Sun, Earth, Moon and Mars were all 4.5 Billion years old, because they were all made at the same time by the Solar Nebula then what we measure is the elements of rocks that form these bodies but 4.5 years ago there was not an instant making of these planets and star..there must have been some sort of process that lead to collecting dust, into rocks and rocks into a planet so the age of the rock should tell us the age of the planet even before it become as such.

Zak Ettamymy

So when we hear that earth and sun are 4.5 billion years old, we are not getting the real information here, earth is made of carbon, silicon and other elements, none of which were made 4.5 billion years ago, they wee assembled 4.5 billion years ago. So in reality earth was assembled 4.5 years ago but everything on earth was made at the Big Bang "if such thing ever happened"

Appendix

This is what we know so far:

Zak Ettamymy

The universe started from nothing and out of nothing it came into existence, a tiny fraction of a second later, the universe experienced a faster than speed of light expansion called inflation, the universe cooled down for the next few seconds and that's when Albert Einstein's' formula E=mc2 kicked in and the first quacks started to form..fast forward a few billion years, stars and galaxies started to shape up including ours, the birth of a star is a slow process which happens inside a cluster of dust called Nebulae and the death of a star happens in a spectacular fashion in phenomenon called super novae, every galaxy has hundreds of billions of stars like our sun inside each galaxy these is a black hole. Black hole is a way a star could end its life when energy and gravity match up and gravity wins. In nutshell a star is a star when energy and gravity are equal, when energy becomes stronger the star dies in an explosion when gravity tips the balance the star becomes a black hole.

The universe is infinite although it started 13.7 years ago, the size of the observable universe is 46 billion years across however the universe could be 100 billion years across or much more, meaning perhaps infinite.

The universe is mostly dark matter which is the reason why it is still intact otherwise it would just scatter to extinction, dark energy is the reason why the universe is still expanding.

Earth is the third rock from a star called Sun, which with other planets represent a system called the solar system; this system occupies a small space in the outer arms of a spiral galaxy called Milky Way, which is part of a cluster of galaxies called the local group.

Earth is 4.5 Billion years old which is the same age as the Sun both were born in a cloud of dust called solar nebula.

All above information are mere prognostications or even prophecies, nothing more than a guess work in a field that unfortunately produced nothing but theories. This is why.

- The universe cannot begin out of nothing (cause and effects)
- We cannot discard the consciousness because it is real (human presence imposed its importance in experiments)
- Earth cannot be the third of the universe age (earth 4.5 universe 13.7)
- Earth cannot be made at the same time as the Sun
- Nebulae cannot create a planet, a star and a galaxy (it is said that Milky Way was also created by a nebula)
- A black hole is a hoax because it does not makes logical sense (it is claimed that time, space and matter stops being at the edge of a black hole)
- The universe is not expanding because nothing on earth and around it is expanding (the expansion of the universe were is everything is flying away from everything has not been proven)

Zak Ettamymy

- Infinity doesn't exist (anything that starts cannot be infinite)
- Inflation is not possible at the rate that Dr. Guth and others claimed
- Parallel universes theory is beyond us because there is no way to see or communicate with the other universe
- The Universe cannot not be 13.7 billion years old while Milky Way galaxy is 13.2billion years old

This is book is not meant to discredit some or any of the world scientists, it s an explanation of why humans are living in total ignorance of their universe.

Humans explored and discovered all of what they called "world" but when it came to the universe the

Zak Ettamymy

scientific community failed to map the universe and explain it to average people in order to have at the very least an understanding of Home.

The baby steps taken so far are courageous and show the dedication of humans as a species that so far sees itself as the guardian of the universe, both morally and physically but inheriting the universe without knowing its borders or even its purpose is an astronomical failure.

I encourage people to keep visiting the planetariums and to watch space documentaries but I also encourage them to challenge any illogical theory put forward, this will spice up the research and perhaps push the scientists to be more forthcoming.

Zak Ettamymy

<u>Sources and Credits</u>

Google Books

Wikipedia

Encyclopedia Britannica

Nova

Youtube

Bob Sagan's Cosmos

What we don't know (the documentary)

Zak Ettamymy

www.ingramcontent.com/pod-product-compliance
Lightning Source LLC
Chambersburg PA
CBHW022121170526
45157CB00004B/1708